CONT

ABOUT THE AUTHORS

Patrick Holford is one of Britain's leading authorities in the field of optimum nutrition. In 1984, he founded the Institute for Optimum Nutrition, an independent centre for the research and practice of nutrition, of which he is co-director. Author of many popular nutrition books including The Whole Health Manual, Vitamin Vitality and The Family Nutrition Workbook, he also contributes to national magazines and frequently appears on radio and television. Currently his time is divided between teaching and training nutritionists, writing and researching.

Dr Philip Barlow is recognised as an authority on environmental health issues and the ways that nutrition can help reduce the adverse effects of pollutants. He has worked as an Environmental Health Officer, advisor to the World Health Organisation, and has lectured on the subject at Aston University , Birmingham. Currently he is a scientific advisor to the Hyperactive Childrens Support Group, Foresight, and the Environmental Medicine Foundation and is Head of the Industrial Services Division in the School of Food and Fisheries at Humberside College of Higher Education.

Acknowledgements
We would like to acknowledge the dedication of scientists such as Professor Derek Bryce-Smith and Dr Neil Ward, and pressure groups like Friends of the Earth for helping expose the serious issues that threaten our health and our survival. A special thanks goes to Anne Pelter for her help in researching, writing and editing sections of this book.

HOW TO PROTECT YOURSELF FROM POLLUTION

Patrick Holford
and
Dr. Philip Barlow

First published in 1990
by ION Press
a division of The Institute for Optimum Nutrition
5 Jerdan Place, London SW6 1BE

© Patrick Holford and Dr.Philip Barlow 1990

Cover Design: QA
Layout: David Simpson

ISBN 1-870976-04-5

Printed and bound in Great Britain by
Brier Press, High Wycombe, Bucks.

 On Recycled Paper

INTRODUCTION

According to top British authorities, if we continue to pollute our environment, the world will become irreversibly damaged within 50 years and inevitably unable to support life. Russian authorities are less optimistic and give us 15 years before it's too late. As the battle to conserve our environment gains momentum, each one of us has a personal battle - to protect ourselves from pollution. In one year the average person breathes in two grams of solid pollution, eats 12 lbs of food additives, has a gallon of pesticides and herbicides sprayed on the fruits and vegetables they eat and receives nitrates, hormone and anti-biotic residues both from water and food. No less than 6,000 new chemicals have been introduced into our food, our homes and the world around us in the last decade alone. These and other pollutants add to the burden our bodies already have to cope with from self-selected harmful substances including alcohol, cigarettes, recreational and medical drugs, free radicals from frying food, methylxanthines in coffee, and a host of naturally occurring toxins present in small amounts even in healthy food.

Although we are equipped with clever mechanisms for detoxifying many harmful substances, these mechanisms are becoming increasingly overloaded. When the total body burden exceeds our ability to detoxify, these substances are integrated into bone, fat, brain and other tissue. When calcium is released for normal body functions so are toxins. When fat is used for energy more toxins are released into the blood - toxins that ultimately affect our brain, nervous system, liver, kidneys and other vital organs.

The effects of pollution are cumulative and are not always seen at the time exposure occurs. It is not easy to prove the link to certain diseases when the pollution may have been slight over decades or have occurred decades earlier, particularly as the origination of most diseases is multifactorial. What we can be pretty certain of are the following negative effects:

Increased lead and aluminium exposure causes a measurable decline in mental performance and behaviour in children and adults.

The level of nitrites and nitrates in water and food contributes to the risk of certain forms of cancer.

The level of free oxidising radicals, both from our polluted air, fried and processed foods, passive smoking and exhaust fumes, coupled with a deficiency in anti-oxidant nutrients, significantly weakens our immune systems and increases our risk of most forms of cancer and heart disease.

The reason why 30 per cent of pregnancies end in miscarriage, and 15 per cent of all babies born have physical defects, is in part the result of increased pollution levels.

These alarming facts represent the tip of the iceberg. Other diseases that have been associated with a high level of pollutants include all forms of arthritis, allergies, candidiasis, M.E., repeated infections, hyperactivity, high blood pressure, asthma, eczema and schizophrenia. 'Minor' symptoms associated with increased body burden of pollutants include lethargy, drowsiness, mood swings, poor concentration, intolerance of alcohol, poor skin, body odour, headaches, nausea, skin rashes, frequent infections and multiple allergies. All of these symptoms can occur for other reasons, such as specific diseases or deficiency in certain essential nutrients. Even when they are the result of increased body pollution these symptoms are far more likely to occur in people who are inadequately nourished. This is because most pollutants are *anti- nutrients* - they do their damage by interfering with the absorption or utilisation of nutrients, or by promoting their excretion. Lead, for example, interferes with calcium and zinc. Many of the symptoms of calcium and zinc deficiency mimic the symptoms of too much lead. What's more, the body's detoxifying mechanisms are themselves greatly enhanced by an optimum intake of essential nutrients, including vitamins, minerals, essential fatty acids and protein.

POLLUTION SOLUTIONS

So how do you protect yourself from pollution? The first step is obvious. Avoid, or reduce as much as possible, your intake of pollutants. In case, by now you're suffering from 'acute pollution paranoia' console yourself with the fact that you cannot avoid all pollution. Harmful substances have always existed even within foods whose overall effect on the body is definitely positive. The best you can do is to minimise your overall exposure; one way to do this is to increase your intake of certain factors that prevent the absorption of pollutants. Copper absorption, for example, is minimised by the presence of zinc. Nitrites are less likely to form the carcinogenic compounds nitrosamines when vitamin C is present. The second step is to boost your natural detoxifying mechanisms. This can be done using diet and supplements.

The first chapter of this book investigates all kinds of pollution - in the air we breathe, the water we drink, the earth in which our food grows, and in food itself - and explains what the problems are and their solutions. The second chapter shows you how to assess where you are most at risk from the effects of pollution. The third chapter shows you how to build your own pollution protection plan, both to protect yourself and your environment.

YOU AND YOUR ENVIRONMENT

There are many levels to your environment. Your first and most important environment is your body. After all, you live in it all the time! Your second environment is your home and workplace.

Your third environment is the world we all live in. As you detoxify your body, and feel the extra energy and clarity that this brings, it is natural to want to live and work in a healthy environment. You will then find yourself becoming increasingly aware of the broader issues and want to make sure that you are doing what you can to reverse the alarming escalation of world pollution and its consequences.

The Different Levels of Your Environment

In reality there is no separation between these different levels of the environment. Everything we do affects everyone else. Every tree burnt on the other side of the world contributes to increasing carbon dioxide levels and changes in the world weather patterns which we all experience more vividly as each year passes. So too does every cigarette smoked and every gallon of fuel burnt. This book explains how you can protect yourself in a way that helps to solve global pollution, not add to it.

CHAPTER ONE

POLLUTION - PROBLEMS AND SOLUTIONS

WATER

Water is, without a doubt, our number one nutrient. The human body consists of no less than 62% water. We need about 2.5 litres of water a day, a litre of which comes from food, the rest from drinks. Water itself, is far more than H_2O. It should provide us with a number of essential elements, the most important of which is calcium. One litre of water can provide us with 103mg of calcium, one fifth of the RDA. Depending on the source, water can also provide significant quantities of potassium and sodium.

In the past fifty years, our drinking water has become contaminated from many different sources. Water now contains nitrates, pesticides and herbicides. Aluminium, chlorine and fluorine are added, and so too are lead and copper from water pipes.

THE ALUMINIUM CONTROVERSY

Recent research has drawn particular attention to the dangers of aluminium in water. Aluminium becomes soluble in water, and absorbable in man when water is more acidic (an increasing problem due to acid rain) or in parts of the country where the water is naturally soft.

To make matters even worse most water authorities add aluminium to the water supply. This is done to clean up the water. Aluminium is a very sociable element and attracts particles in the water thereby cleaning it. The same objective could be achieved by filtration, although the costs are higher, or by adding in iron in the form of ferrous sulphate, which would be far safer since, if anything, there is a general deficiency in iron, especially in women. Currently 100,000 tons of aluminium is added to the British water supply each year, at a cost of £7 million. Most water authorities now deliver water with levels often up to three times the permitted safe level set by the EC.

Water is not the only source of aluminium. It is also high in processed cheese, drugs including aspirin and antacids, antiseptic medication, deodorants, toothpaste, baking powder, some table salt, dry milk, instant coffee, tea bags, and is added to foods (E173) as an emulsifying, bleaching or anti-caking agent. Aluminium also contaminates food from aluminium foil, packaging and pots and pans.

ALUMINIUM, MEMORY LOSS AND HEALTH

Aluminium was first associated with memory loss and Alzheimer's disease, a form of senile dementia, in the 1960's, an association that has been strengthened by the discovery that the plaques that form in the brain of Alzheimer's patients are loaded with aluminium. In Southern Norway where aluminium levels are high in lakes, due to acid rain, the areas with the highest aluminium levels have the highest incidence of senile dementia. Alzheimer's disease is very common in the elderly, affecting no less than 22 in every 100 people above the age of 80.

Aluminium interferes with the minerals calcium, iron and zinc. Any deficiency in these minerals leads to a much greater absorption of aluminium. For this reason anaemic people, who are iron deficient, usually have raised aluminium levels. Aluminium excess has now been associated with behavioural problems, nervous system disorders, extreme tiredness, and skin problems, as well as disorders of calcium balance including osteoporosis. One study by Dr Neil Ward and co-workers at the University of Surrey has shown a close correlation between low birth weight babies, excess aluminium and zinc deficiency.

The soya bean is naturally high in aluminium and, according to the American Committee on Nutrition, highly processed infant formulas based on soya milk are not recommended to low birth weight babies, premature babies or babies with kidney problems, all of whom are more likely to be zinc deficient. However, if your zinc, iron and calcium status are good there is no reason to avoid soya products.

The danger of adding aluminium to water is well illustrated by one of four recent incidents when a water authority added excess aluminium by accident directly into the water supply. This incident in Cornwall involving the South West water authority resulted in levels initially reported to be up to 109 parts (the EEC allows 0.2 parts), although six months later South West water authority admitted that they had measured levels up to 600 parts! As the water was as acidic as vinegar, it became packed with aluminium. The high acid level also released high levels of copper and lead from water pipes. A study was conducted on 21 children who had been exposed to contaminated water for up to five days while on holiday in Cornwall. It revealed that, six months later, sixteen still had skin rashes and eczema, four had bowel disorders, twelve had chronic fatigue, four were depressed, and six had developed language and speech problems. Twenty thousand people were poisoned from this particular incident, and an equivalent number of fish died in local rivers.

Despite these incidents, most water authorities continue to add aluminium to water and continue to exceed EC safety levels - a situation that is likely to get worse, not better, with privatisation.

The best way to protect yourself from aluminium is to minimise avoidable sources, filter your water, and ensure an adequate intake of calcium, iron and zinc.

THE QUESTION OF FLUORIDATION

Fluoride is another element added to water for dubious reasons. Like aluminium it upsets the chemistry of the body and interferes with calcium, magnesium, iron and zinc. It is not essential and acts as a poison even in low doses. There is evidence that, at levels up to 1ppm (part per million) in water it may prevent tooth decay. However, at a level of 2ppm it causes *fluorosis* , staining and mottling of teeth. We also receive fluoride from dental treatments and toothpaste. In areas where water is fluoridated processed foods, for example cheese, contain significant amounts of fluoride. Therefore it is impossible to ensure that some people do not receive amounts of fluoride that exceed those considered safe. These safe levels themselves depend on other factors. Fluoride, like aluminium, is much more readily absorbed from an acid medium, namely soft water, or if calcium, magnesium, zinc or iron levels are low.

Despite all these uncontrollable factors, the British government passed a law giving health authorities the right to add fluoride into water for the sole purpose of preventing tooth decay. Apart from introducing a potential health risk, fluoridation represents enforced medication to prevent a problem which is acknowledged to be the result of faulty nutrition. What's more, in those countries that have taken up fluoridation recently, decline in the incidence of tooth decay has been far less than expected.

Meanwhile, in other countries which began fluoridation many years ago, and have experienced some of the dangers, fluoridation is no longer popular. Austria, Belgium, Denmark, France, Greece, Holland, Italy, Luxembourg, Norway, Spain, Sweden, West Germany and Yugoslavia have all rejected fluoridation and currently less than 8 per cent of people in Britain drink fluoridated water.

Fluoride inhibits enzymes in the body and in large amounts poses serious health problems. In Chile, fluoridation was discontinued in 1977 because of an increased incidence of child mortality in fluoridated areas. Research showed that malnourished children were more susceptible to fluoride toxicity. There is also some suggestion that excess fluoride intake may increase susceptibility to cancer. Whatever the truth to these more alarming possibilities, there is no doubt that fluoridating water is ethically unsound.

The best way to protect against a potential excess of fluoride if you live in a fluoridated region is to filter your water and ensure a good intake of calcium, and also magnesium, zinc and iron.

HOW SAFE IS CHLORINE?

Chlorine is the standard treatment for removing bacteria from water. Despite its widespread use for almost a century, chlorine, or rather its by-products, are now under investigation following American research which found 'a highly significant relationship' between chloroform and cancers, particularly cancers of the bladder, colon and rectum. Chloroform, which is part of a group of substances called trihalomethanes (THMs for short), can be created from chlorine when it reacts with peat and other organic material found in water. The EC safety level for THMs is 100mcg/l although West Germany has adopted an upper limit of 25mcg/l in view of the potential toxicity. Needless to say many water authorities in Britain exceed the EC safety level, in order to disinfect our water due to the high levels of sewarage contamination.

THE NITRATE SCANDAL

All plants need nitrogen compounds to grow. They are an important constituent of protein, which is normally taken up from the soil. By feeding plants more nitrates it is possible to speed up plant growth, and hence make more profit. If the nitrates are incorporated into protein by the plant this does not present a problem, but if they're not, the nitrates are passed onto us when we eat the plants, or they may leach straight into the water supply. Governments and the World Health Organisation have known for many years that nitrates and nitrites are potentially toxic, and set limits for water and food levels in 1970. Currently 1.7 million people in Britain are consuming water containing nitrates above this safety limit. The problem is especially severe in arable areas such as Norfolk and East Anglia.

Once inside the body, nitrates can be converted to nitrites and then combine with amines, present in almost all foods, to form *nitrosamines*, which are highly carcinogenic compounds.

Nitrates are also high in vegetables grown with nitrate based fertilisers. Nitrates are also added to cured meats such as ham, sausages, bacon and pies. 70% of our intake comes from vegetables, 21% from water and 6% from meat. Nitrites are also produced in the mouth by bacteria - another reason to brush and floss your teeth regularly. But the main protection is to drink filtered or bottled water and eat organically grown produce wherever possible. Also, meat products preserved with nitrates or nitrites should be avoided. In fact, it is best to keep protein intake adequate, but not excessive. Vegetarian diets, if well-balanced, are generally protein sufficient.

The best way to protect against nitrates and nitrites is to make sure your diet is rich in vitamin C and supplement at least 1,000mg a day. Vitamin C inhibits nitrosamine formation, and hence it is ideal to have a dietary or supplementary source of vitamin C with each meal. But the only long term solution is to persuade farmers to stop using nitrate based fertilisers and switch to organic farming methods.

WATER - WHAT ARE THE OPTIONS ?

Most tap water in Britain now exceeds the safety level for one or more pollutants. This is a situation that is likely to get worse, not better, with privatisation. It is wise to reduce your consumption of tap water and to ensure adequate nutrition to minimise the effects of pollutants in water. Other options include bottled water, distilled water and water filters.

Many bottled waters come from underground or glacial sources and are usually very pure. However, it is not a practical long term solution to the problem of poor quality water, costs of transport and

packaging being huge. Bottled water costs from 30-70p per litre.
Excluding their initial costs of about £10, jug filters provide water at a cost of 3p-7p per litre, due to the cartridge having to be replaced at least monthly. Distilling water costs about 6p of electricity per litre and is a slow process. The unit itself costs £200-400. Plumbed in filters cost about £60-250, they generally last about 2 years - replacement cartridges cost from £30-150, giving water at a cost of about 1p per litre. Reverse osmosis units cost from £400-1200.

WATER FILTERS - HOW DO THEY WORK?

The simplest types of water filter are the plastic jug varieties. Most of these contain activated carbon and ion exchange resins. Activated carbon has a large surface area which attracts impurities from the water, such as chlorine and organic compounds. Most activated carbon has some silver added to it. This inhibits bacterial growth, but may leach into the water. The ion exchange resins attract heavy metals like lead but also essential minerals like calcium and magnesium.

Plumbed in filters supply a separate tap with filtered water and once again usually contain activated carbon. Some carbon type filters can become breeding grounds for bacteria if not frequently changed.

Reverse osmosis filters remove asbestos, bacteria, viruses, fluoride, aluminium, heavy metals, salts, nitrates and minerals (including calcium and magnesium). Up to four times the amount of water produced for drinking is rejected and wasted; also the water is said to be rather acidic.

Distilled water is pure but insipid and a lot of energy is used in its production. Distilled water is the softest available. Soft water has been linked in many studies to ill-health. The lack of calcium, for example, is not good for the health of teeth, bones, heart and blood vessels. While theoretically individuals could make good their mineral levels with supplements, in practise this might be difficult to implement across the population. Also, distillation uses large amounts of energy in it's production and thus is not advisable when we should all be trying to cut down on our energy consumption.

Water softeners allow water to pass over a resin that holds the magnesium and calcium ions - but releases sodium ions. Water regulations forbid the connection of water softeners to drinking water taps - they may be good for your washing machine, but they're not good for you.

If you do not use a filter it is a wise precaution to draw off water for several minutes in the morning to avoid water that has been standing in lead pipes overnight. Do not drink hot water directly from the tap, as heat encourages copper and lead from pipes and aluminium from the immersion tank to dissolve in the water.

Your Public Health Department may carry out a test on your water supply, free of charge. If you do decide to replace the lead piping in your house, then the Water Authority must also replace the pipes leading to your home.

 AVOID

• Drink filtered or bottled water.
• If you don't, run off water for a couple of minutes in the morning before drinking.
• Don't drink water from the hot tap.
• If your water is fluoridated avoid other sources of fluoride.

 **PROTECT**

• Ensure that your intake of calcium, magnesium, iron and zinc is more than adequate. This protects you against aluminium, fluoride and chloride.
• Supplement your diet with 1,000mg of vitamin C and eat foods rich in vitamin C. This protects against nitrosamine formation.
• Brush and floss your teeth regularly. This also protects against nitrosamine formation.
• If you have a water filter change the filter element according to the manufacturers recommendations.

AIR

Air should be a mixture of 78% nitrogen, 21% oxygen and less than 1% carbon dioxide. However, the air we breathe these days contains a whole range of other gases, vapours and particles derived from either natural sources, such as volcanoes, or from man's activities.

Air pollution is the discharge into the air of excessive amounts of unwanted substances. This includes abnormal amounts of some of the natural components of air, such as carbon dioxide. Such changes in the composition of the air can have serious effects on our health. An example is the amount of carbon dioxide produced from burning fossil fuels which has created the alarming spectacle of global warming, resulting in calamitous weather changes.

The problems of air pollution are greatest in cities. The exhaust fumes from cars are a particular problem as they contain carbon monoxide, lead, oxides of nitrogen and many hydrocarbons.

WE BREATHE IN WHAT PLANTS BREATHE OUT

There is a balance between the plant and animal kingdoms. The plants take up the waste carbon dioxide we exhale, plus water to produce carbohydrates. One of the consequences of reducing the amount of trees (ie plants) to meet demands for fuels, paper making and grazing pastures for beef, is that less carbon dioxide is taken up. This builds up in the environment, while oxygen levels decline as less is given off by plants due to their destruction.

In addition to air pollutants in the general environment, we are exposed to tobacco smoke, contaminated air at work and from cooking and heating appliances. Let us now examine some of the specific air pollutants and see how you can protect yourself from their ill effects.

LEAD OR HEALTH

Lead is a powerful nerve toxin which damages the brain and nervous system. The early signs of lead poisoning are altered behaviour and intelligence. Symptoms can include headaches, tiredness, insomnia, hyperactivity, depression, irritability, aggressive outbursts, frequent colds and a loss of appetite. Children are most susceptible to the effects of lead and numerous studies have shown that low levels of lead, around 13mcg/dl in blood, significantly reduce intelligence and behaviour. In 1980 the average level of lead in inner cities in Britain was 12.8mcg/dl, leading to the appalling conclusion that lead was effecting the intelligence of most city children.

The majority of lead in air originates from exhaust fumes. Car exhaust accounts for ten times as much as that released by industry. Lead levels in the UK fell by over 50% between 1985 and 1988 with the introduction of unleaded petrol. However we are still exposed to levels 500 to 1,000 times higher than our ancestors. Currently unleaded petrol still only accounts for 26% of the market.

HEAVY METAL KIDS

Children, due to their height and their exposure to dust when playing, are most at risk from lead exposure. As well as keeping children away from potential sources of pollution, it's important to keep them well nourished to help detoxify the pollution they are inevitably exposed to. Calcium, zinc, vitamin C and vitamin B1 effectively reduce the toxicity of lead both in children and adults. Diets low in zinc and calcium, or high in alcohol, cause the body to absorb much more lead.

The concentration of air lead inside cars, even with the windows closed, can be about 75% of the concentration outside. So it is best to avoid spending long periods of time in a car on busy roads or motorways, especially for children.

The food you select also makes a difference. Food grown or sold close to busy roads may contain lead fall-out. Outer leaves of green vegetables should be discarded and the edible portions well washed before consumption. Adding a small amount of vinegar to the water you wash your vegetables in helps remove lead. Washing will remove up to 90% of the lead, but the final 10% is extremely difficult to remove.

Another element associated with mental and behavioural problems is *cadmium*. The discovery of high levels of cadmium in smokers provides another reason to stop. Professor Bryce-Smith from Reading University demonstrated the damaging effects of these elements in an extraordinary piece of research. He analysed mineral levels in the placenta of 'normal' mothers, immediately after birth. He was able to predict both the birthweight of the newborn babies and their head circumference (a factor associated with mental retardation) from the levels of lead, cadmium and zinc. The higher the lead and cadmium levels and the lower the zinc in the mother, the lower the birth weight and head circumference!

THE DANGERS OF PASSIVE SMOKING

Smokers are at greater risk from airborne pollutants for two main reasons. Firstly the air/smoke mixture contains a high proportion of carbon monoxide and hydrocarbons, many of which are known to be carcinogenic. Second, cigarette smoke depresses the activity of the small hairs in the respiratory tract whose job it is to sweep out any particles that are inhaled. So smoking not only introduces more pollutants, but also depresses the body defence mechanisms.

Apart from the inhalation of tobacco smoke by the person smoking, non-smokers are exposed to pollution in a smoke filled room. There may well be less oxygen and more carbon monoxide. This can lead to a decrease in brain activity resulting in a reduced efficiency. Passive smoking has been linked to a variety of respiratory diseases including lung cancer and a greater risk for heart disease. More than 20 studies have been conducted comparing death rates due to lung cancer in woman married to smokers with those married to non-smokers. These point to about a 30% increase in risk. A recent study concluded that young children have a greater risk of respiratory problems if their parents smoke.

If you smoke or are exposed to someone else's cigarette smoke it is important to include in your diet plenty of fresh foods rich in *anti-oxidant nutrients* namely vitamins A, C and E and the minerals selenium and zinc. We estimate that maximum protection is effected by taking in 40mg of vitamin C per cigarette.

ACID RAIN AND THE OZONE LAYER

Industrial pollution gives rise to gases that acidify water particles in the air. One of the most concerning is *sulphur dioxide*, produced by burning oil or coal. In high concentrations it can cause serious damage, irritating the eyes and lungs and causing severe respiratory distress. At low concentrations it is dealt with by the lungs and changed to sulphate, which is harmless.

When this *acid rain* falls it damages plant life and picks up more pollutants in the soil which are then washed into streams and rivers which may subsequently be used for drinking water.

Ozone is formed in air by the reaction of hydrocarbons with oxides of nitrogen in the presence of sunlight. The sunlight is necessary to provide the energy for the chemical reaction to take place. Thus ozone formation is a special problem in areas of high sunlight. In the UK it is primarily a problem in summer haze. Ozone forms at street level and is especially a problem in areas of high traffic density as both hydrocarbons and oxides of nitrogen are emitted from car exhausts. Long term exposure to ozone causes a thickening of lung tissue and a decrease in lung efficiency.

Recent research also suggests that if a person is exposed to more than one air pollutant, for example ozone and sulphur dioxide, the combined effect is greater than the sum of the two individual effects. The long term effect on people exposed to a whole range of pollutants is just not known. However, an association between air pollutants such as ozone, sulphur dioxide and nitrogen oxides and asthma has been shown, putting people with respiratory tract problems at greater risk from air pollution. Once again, a good intake of anti-oxidant nutrients helps protect you from the harmful effects of both ozone and other air-bourne pollution.

NO CFC'S PLEASE

Chlorofluorocarbons (CFC's) are used in many aerosol products and refrigeration systems. They are capable of destroying the upper atmosphere ozone layer. This does present a health hazard in that the ozone layer at high levels (unlike the ozone formed at low levels from car exhaust) has a screening effect on the ultra-violet radiation from the sun. If we are exposed to too much UV light there is a risk of developing skin cancer - as there is if we sunbathe for too long without taking the proper precautions. Always try and   purchase 'ozone friendly' aerosols.

LIVING WITH RADIATION

Radiation doesn't just occur from man made sources such as nuclear power generation or X-rays. We are all exposed to radiation from the sun and deep space. There are even naturally occurring radioactive materials in air, food and water. It is estimated that the average person in the UK receives about 87% of their annual radiation dose from natural sources and 11.5% from medical X-rays. The remaining 1.5% comes from artificial non-medical sources such as that from nuclear power generation - unless something goes wrong, such as Chernobyl.

The term radiation is wide and includes such things as light waves, radio waves and microwaves. However it is most often restricted to mean *ionising radiation*, that is radiation that produces free radicals, which, in turn, can damage living organisms. While the initial effects of radiation are usually invisible the long term consequences of excessive exposure can include leukaemia or other kinds of cancer, infertility, eye cataracts or skin damage.

RADON

One of the largest 'natural' sources of radiation is the gas *radon*. This material is an important cause of human exposure to radiation, most of which occurs indoors. Radon gas is part of the decay chain of uranium to lead. Uranium and it's decay product radium are found naturally in rocks and soil and are found in building materials such as wood, bricks and concrete. When the decay products are inhaled (contained in small dust particles), the radioactive particles settle in the lungs and irradiate intensely at close range for years.

In the open air any radon is mixed and diluted with air and quickly dispersed. However, inside buildings, radon is released from building materials and from the ground, and are then inhaled by the occupants.

Recent surveys carried out in the UK suggest that residents in some parts of the country are at greater risk than others. The southwest seems to be the most affected area but there are other local 'hot spots' which have been identified. Areas with high granite are the most affected. In some cases the radiation dose from radon accounts for over 50% of the total natural radiation. At this level it is suggested that it could be responsible for about 500 deaths from cancer per year. If you live in a high granite area contact your local Environmental Health Officer to find out what your radon exposure is likely to be. When radon levels are high certain building materials are better to use than others. Adequate ventilation under floor boards is also important, to remove the radioactive compounds before they can build up to significant levels.

THE DANGERS OF WATCHING TELEVISION

Many people these days use a VDU screen at work or possibly at home. VDU screens generate an electrostatic field which attracts dust particles towards the screen or the operators face. This can lead to irritation of the eyes, runny noses and dry and itchy skin rashes. This effect reduces rapidly with distance so there is not normally any problem watching the most common type of VDU (ie the television screen) unless a person sits right up close to the set. Children should therefore be discouraged from sitting close up to the television.

When operating word processors, for example, a person sits much closer to the screen and prolonged exposure might cause health problems. One way of reducing the problem is to fit an electrically grounded filter in front of the screen. Such devices dramatically reduce the movement of particles towards the operator. However they do not greatly screen out the radiation emitted from the screen. This radiation however is not ionising radiation and is not thought to be a health hazard. One other concern as regards

VDU screens is the magnetic fields created around them. Currently it is not known if these fields affect the body adversely. However, it is recommended that pregnant women limit their exposure to such radiation fields in case they prove to be harmful.

PROTECTING AGAINST RADIATION

Radioactive elements can be taken up by the body and incorporated into body tissues where they continue to emit radiation. An example is radioactive iodine which is taken up by the thyroid gland. By ensuring a more than adequate intake of minerals, in this case iodine, tissues are saturated with non-radioactive elements and are therefore protected. The anti-oxidant nutrients A, C and E and the minerals zinc and selenium are helps protect against the damage caused by ionising radiation.

FREE RADICALS - ARE DAMAGING YOUR HEALTH

Radiation is one cause of an especially dangerous group of unstable substances called *free oxidising radicals*. Unlike most atoms or groups of atoms, called molecules, free radicals have an uneven electrical charge. Like a wanton bachelor looking for a mate, in order to complete themselves they attack other atoms to steal a charged particle. Although they then become stable, free radicals set up a chain reaction of damage, creating more free radicals. Cell walls and DNA, the code within cells, are most prone to free radical damage. This damage can either result in the cell dying, or at least becoming faulty. This is how cancer starts. Free radical damage is intimately associated with most forms of cancer, heart disease and premature ageing.

Heat, as well as radiation, increases our exposure to free radicals. For example, ozone is a free radical. Frying food in vegetable oils is another significant source of free radicals. The high temperature, coupled with lots of double bonds (which free radicals love to attack) found in polyunsaturated oils, creates many free radicals. Ironically frying with butter, which has no double bonds, is better from the free radical point of view. But frying is best kept to a minimum anyway. Anything burnt is a source of free radicals - even toast, barbecued or smoked food, cigarettes and car fumes.

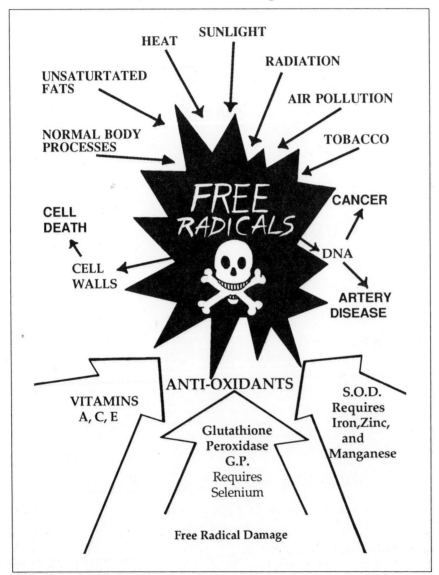

But not all sources of free radicals are avoidable, or even desirable to avoid. For example, the act of using oxygen together with glucose, to make energy in every one of our body cells gives rise to free radicals. Some immune cells even make free radicals to destroy viruses and other alien invaders. We can keep these dangerous molecules in check with anti-oxidant nutrients, namely vitamin A, both as retinol, the animal form, and beta carotene, the vegetable form, vitamin C and vitamin E. The minerals zinc, selenium, iron and manganese are involved in anti-oxidant enzymes, as are the amino acids glutathione and cysteine. Nature also uses these nutrients to protect foods particularly prone to free radical damage. Seeds, rich in poly-unsaturated oils, are a good source of vitamin E.

INDOOR AIR POLLUTION

The improvement in housing conditions with double glazing and the great reduction in the number of open fires has led to a reduction in indoor ventilation. In many modern homes the air exchange rate may be as low as once in every 10 hours. Therefore any pollutant in the house may build up to undesirable concentrations or give a low level of exposure to the inhabitants over a long period of time. Examples of indoor pollutants include carbon monoxide, carbon dioxide, sulphur dioxide and oxides of nitrogen arising from the burning of cooking and heating fuels, *formaldehyde* from such sources as cavity wall insulation, carpet backings and underlay. Formaldehyde is known to be mutagenic (ie can affect the genetic material and cause malformations in the next generation) and carcinogenic certainly in animals and probably in man. It does cause allergic reactions in some people and is an extremely irritating and unpleasant gas. According to recent NASA research the spider plant can help turn formaldehyde into a non-toxic substance!

As well as ensuring adequate ventilation a number of air filtration systems are now available. The best kind contain mechanical filters for large particles and ionisers, which electrically charge small particles which are then attracted to earthed plates. As well as removing pollutants from the air there is evidence that an increase in negatively charged ions, produced by ionisers, is good for your health. This may explain why some people feel better after a thunder storm or by the sea, where the air is more negatively charged.

 AVOID

• Avoid buying unwrapped fruit and vegetables exposed to street traffic that do not require peeling.
• Wash all fruit and vegetables, preferably in a bowl of water with two tablespoons of vinegar added. This acidifies the water and helps remove heavy metals.
• Discard the outer leaves of cabbage, lettuce etc.
• Do not exercise by busy roads. In cities, the lead level is highest during overcast days, and lowest on a clear day after rain.
• Avoid spending long times in heavy traffic.
• Avoid polluted areas on hot, sunny days.
• Don't smoke and, where possible, avoid smoke places.
• Avoid using aerosols.
• Minimise your intake of fried foods, burnt or barbecued foods.

 PROTECT

• Make sure your diet provides plenty of calcium, zinc and vitamin C which are heavy metal antagonists. That means eating lots of nuts, seeds, green leafy vegetables, fresh fruit, and wheatgerm.
• Eat foods high in pectin, which is also protective. This includes apples, bananas and carrots.
• Minimise alcohol which increases the absorption of lead.
• Supplement your diet with at least 1,000mg of vitamin C, 200mg of calcium and 10mg of zinc. Double this amount if you live in a city.
• Supplement your diet with anti-oxidant nutrients such as vitamins A, C, E, zinc and selenium.
• Get an ioniser and make sure your home and workplace are well ventilated.

EARTH

When you buy food prepackaged in the supermarket it's hard to make the connection that food comes from the soil, and that the condition of the soil is perhaps the most vital factor in determining how healthy the food will be. Minerals in the soil pass into plants, and hence into us, sometimes via meat, fish, eggs or cheese. Protein is made by the vegetable kingdom by combining nitrogen from the soil with other elements. Water, our most essential nutrient, comes to us via the soil.

Modern farming methods have done much to mess up the soil. Overfarming has led to demineralisation of the soil. Consequently, food mineral concentrations are declining and mineral deficiency is becoming widespread. The use of herbicides, pesticides and nitrates have acidified the soil, which makes the minerals that are present less available to plants. As well as killing undesirable 'pests', beneficial insects and micro-organisms, such as worms which help to aerate the soil, are also killed. The net result is that soil is becoming increasingly unable to support plant life, leading farmers to depend even more on chemical methods of farming. The only viable long-term answer is to restore health to the soil through remineralisation and organic farming.

PESTICIDES

Pesticides are a group of chemical substances designed to kill specific organisms, including insects, rodents, weeds and fungi. Pesticides are therefore poisonous materials designed to be toxic to one group of organisms, but likely be toxic to others including man. 26 million kilograms of pesticides are used each year here in the UK. It is now estimated that up to a gallon of pesticides is sprayed onto the fruit and vegetables a person may eat in a year. They are used in the street, in parks, workplaces and homes and affect everyone to some degree. Pesticides contaminate our food, the air we breathe and the water we drink.

Pesticides come as many different types including *metal containing compounds* (these can contain arsenic, copper, mercury and tin along with many other metals); *phosphorous containing compounds* mainly used as insecticides because they act on the central nervous system; *nitrogen containing compounds* mainly used as herbicides as well as insecticides; *halogen containing compounds* (ie containing chlorine, bromine or iodine). Halogen compounds are generally of lower acute toxicity than the phosphorous ones but do pose environmental problems because of their persistence in the environment. Examples of this class of pesticides are DDT, *Aldrin and Dieldrin*. The 'drins' are highly chlorinated compounds which are extremely persistent and can only be broken down at high temperatures.

PESTICIDES AND HEALTH

There is little doubt that over the years pesticides have contributed greatly to the improvement of human and animal health. When we buy food we like it to be free from mould or evidence of pest damage. If food is affected by pests not only does it look unattractive but it might be contaminated with bacteria or parasites which could be harmful to health. Thus pesticides can play an important role in food hygiene. They prevent contamination being spread by rats, flies and other insects while food is in store, on display or awaiting use. It is estimated that without the use of pesticides about one-third of the world's food crops would be lost to pests.

THE DANGERS OF PESTICIDES

But with today's chemical farming even the old adage that 'an apple a day keeps the doctor away' must be questioned. For the caterpillar, one brief journey across the average apple is enough to kill it. But what about us? In 1984 over 3,000 incidents of illness resulting from pesticide poisoning were recorded - including one death.

A survey by the Ministry of Agriculture, Fisheries and Foods found that between 89% and 99% of all fresh fruit, cereals and vegetables is sprayed with pesticides. That means that most meat and milk is contaminated by pesticides used in the animal's food.

POLLUTION - PROBLEMS AND SOLUTIONS

Although few proper surveys have been carried out to find the size of the problem, the Association for Public Analysts is gravely concerned. The association randomly tested 305 fruits. 31 of these samples contained levels of pesticides above the levels deemed safe, and a further 72 samples showed lower pesticide levels. Some fruits particularly strawberries, raspberries, grapes and tomatoes had measurable levels of at least six different pesticides!

Even more alarming is the fact that some of the pesticides found are not 'cleared for use' in Britain, while others, although permitted, are banned in many other countries because of their known damage to health in promoting cancer and causing birth defects. DDT, for example, was officially banned in 1986. However a further survey by the Public Analysts found that, out of 293 samples of different vegetables, 10% contained DDT!

There are many dangers in our current overuse of pesticides. Most crops are treated more than once and some food products are sprayed many times during their growing cycle, yet the possible toxic effects of these combinations of pesticides has not been fully investigated. It is likely that some do potentiate the toxicity of others. Also, many pesticides are applied directly to the edible parts of plants after harvesting. These residues will therefore be eaten by us. Other pesticides are highly persistent in the environment and once released just pass along the food chain until they reach us. Pesticides also acidify the soil and destroy beneficial micro-organisms, leading to poor soil and poor crops.

Pesticides such as DDT or Dieldrin have already caused a drastic decline in the British otter population and birds of prey, such as the sparrow hawk. Both these pesticides are now banned from use in the UK but it is likely that traces of them will be found in the environment for many years to come. Pesticides have also contaminated most rivers and their inhabitants - fish.

Pesticide levels in drinking water now often exceed EC safety limits especially in East Anglia, the East Midlands and London and the Home Counties. It is thought that some of these substances are carcinogenic; in addition the long term effects of ingesting these trace amounts of poisons are not known. It is strongly suspected that they may well be involved in the lowering of the body's immune system response.

Many of the incidences of pesticide poisoning are the result of a short-term concentrated exposure. The symptoms can include eye and skin irritation, headache, central nervous system disturbance and breathing distress. Perhaps even more concerning are those pesticides that damage health in the long term, as low levels accumulate in the body.

An example is the chemical *daminozide,* known as *Alar,* which is sprayed onto ripening apples in order to make them more attractive. In animals it produces cancer. Who knows what effect it is having on children? Another undesirable is *2.4.5.T,* which can contain traces of *Dioxin..* The types of body response can range from cancer, liver function disturbance and kidney disorders to central nervous system disturbance and next generation birth defects. Alar has recently been withdrawn. Moves to ban other potentially unsafe chemicals need our urgent public support.

While washing and removal of outer leaves etc. will help to reduce residue levels they cannot be relied on to totally eliminate pesticides. The best precaution is to eat organic.

CHEMICAL FREE FOOD

The increased demand for natural and additive free foods is catalysing a return to traditional methods of growing foods. Currently about 1% of farmers are producing organically grown food. Interestingly, a good deal of the organically grown food is imported. At present there is no formal legislation on organic farming and some produce sold under the title 'organically grown' still contains pesticide residues.

Organic farming aims to grow foods without recourse to the use of pesticides, or chemical fertilisers such as nitrates. It takes a while to clear the soil of these residues and restore the normal balance of micro-organisms. One step on from organic farming is to restore the mineral balance in the soil by adding gravel dust, literally crushed rocks, to the soil. Manure and compost can then be used to return goodness to the soil. In this way the soil and the plants that grow in it are kept healthy. The healthier the plants, the more resistant they are to pests and disease and the richer they are in nutrients.

WHAT'S SO GOOD ABOUT ORGANIC?

Organic produce, because it is not treated with chemicals to delay ripening, prolong shelf life or preserve colour, consequently doesn't always look so good. But don't be fooled by appearances. It's much better for you. Cattle fed on organic produce consistently produce higher milk yields. Rabbits have higher pregnancy rates, larger litters and healthier offspring.

Organic produce has a higher dry weight value. In other words non-organic produce contains more water, as well as pesticides and nitrates from NPK fertilizer (nitrogen-phosphorous-potassium). The extra nitrates make copper less available to the plant, while phosphorous makes zinc, and copper less absorbable. Potassium can make calcium, magnesium and boron less available too. This is consistent with comparisons of the nutrient content of organic versus non-organic produce which shows increased levels of protein, vitamin C, potassium, magnesium, calcium, copper, zinc and manganese.

COMPARISON OF ORGANIC vs NON-ORGANIC PRODUCE	
Component	Mean % increase in organic vs non-organic produce
Dry matter	+26%
Potassium	+13%
Calcium	+56%
Magnesium	+49%
Iron	+290%
Copper	+34%
Manganese	+28%
Protein	+12%
Essential amino acids	+35%
Nitrates	-69%
Phosphorous	-6%

Source: Biological Husbandry: A Scientific Approach to Organic Farming ed.B.Stonehouse, Butterworths 1981.

Organic produce is currently more expensive, however it is not necessarily more expensive in terms of the nutrients that it contains. After all, since most organic produce has at least 20% more dry matter one would expect to pay 20% more for it. As it becomes more available and more in demand, the cost will come down. Since organic produce does not resort to using chemicals to delay ripening it is best to choose foods in season. Organic produce actually stores better than non-organic produce that has not been specifically treated with chemicals to increase shelf life.

 AVOID

• Select organic fruit and vegetables wherever possible.
• Wash or peel non-organic produce.
• Choose fruits and vegetables in season. This means that your exposure to the chemicals used to delay ripening and prolong shelf life will be limited.

 PROTECT

• Supplement your diet with anti-oxidant vitamins such as A, C and E and the minerals zinc and selenium since the detoxification of many pesticides involves these nutrients.

FOOD

However healthy food is to start with, there are many potentially polluting processes it can go through before it reaches our mouths. Many chemical food additives are still used. A number of chemicals are purposely added to our food to change its colour, preserve it longer, prevent rancidity and keep fats emulsified and foods stable. Most of them are synthetic compounds, some with known negative health effects. But the important point is that we really don't know what the long-term consequences of consuming relatively large amounts of chemical additives is, nor do we know if they become more toxic in combination.

It is therefore best to avoid all chemical food additives, and therefore E numbers, with a few notable exceptions. The following E numbers are just natural substances. E101 is vitamin B2, E160 is carotene, vitamin A. E300-304 are vitamin C, while E306-309 are vitamin E compounds used as preservatives. Lecithin is E322, used as an emulsifier, while E375 (vitamin B3) and E440 (pectin) are used as stabilisers.

THE DANGERS OF FOOD PROCESSING

The amount of actual processing and cooking the food is exposed to may kill off or remove vital nutrients that protect us against pollution. In the case of refining flour, cadmium remains, but zinc, which is rich in the bran and protects against cadmium, is lost, making the natural cadmium content more concerning.

Processing involving high temperatures, especially frying, is likely to increase oxidation of the food, leading to free radical formation. The longer foods are stored - especially if exposed to light, heat or air - the more oxidation can occur. For this reason canning, freezing, freeze-drying and bottling helps to preserve foods. Supplements are also prone to oxidation and are therefore best stored in amber glass bottles, in a cool dry place. Cold pressed oils and wheatgerm are best stored in the fridge.

MICROWAVES

Microwave ovens, producing non-ionising radiation, do not create free radicals. The microwaves vibrate water particles in the food which then cooks it. Since there is little water loss there is little loss of water soluble B vitamins and minerals. There is some suggestion that microwaves may have detrimental effects on the cardiovascular and nervous system. However modern microwave ovens do not 'leak' microwaves when in use and should not pose any health hazard. Food does not appear to be altered any more significantly than by other forms of cooking.

POLLUTION FREE PACKAGING

All foods packaged in aluminium are best avoided. If the food is very acidic the aluminium is likely to leach into the food. Some fruit juice containers are lined with aluminium, which is not recommended; a viable alternative is plastic. Canned food used to be frowned upon due to the use of lead solder, but most canned foods no longer use a lead seal. Many paper products are bleached with chlorine, which can give rise to a toxin called dioxin. The US Environmental Protection Agency has found levels of dioxin 1,000 times higher than the maximum daily limits in cartons of milk for school children. In animals, dioxin has been linked to a variety of immune system diseases. When foods are in direct contact with paper products it is best to check that they're dioxin free. For example, some herb tea manufacturers use dioxin free teabags, others don't. If in doubt, check with the manufacturers.

Waste paper alone accounts for 130 million trees each year, therefore, when there is a choice, choose products packaged in recycled paper or cardboard. Glass bottles are preferable to plastic - again because they can be recycled.

FOOD IRRADIATION

There now seems little doubt that the UK government will legalise the process of food irradiation. To date it has been banned in this country but is widely used in a number of other European countries, along with many other parts of the world. People who have travelled abroad are quite likely to have consumed food that has been irradiated!

The irradiation of food is a technological process aimed primarily at improving and extending its shelf life. The application of ionising radiation is intended to reduce or eliminate bacteria responsible for food poisoning, along with parasites and insects. The process also prevents the sprouting of vegetables such as potatoes and onions during any prolonged storage.

Irradiated fish, fruit and vegetables generally remain 'fresh' and the physical state of products unchanged. If there are all these advantages for irradiation as a method of food preservation, why are many people not happy about its introduction and why are some supermarket chains deciding not to stock irradiated food?

If the process is to be accepted the following matters need to be answered satisfactorily. How is the nutrition quality of the food affected? Are any toxic compounds produced during or after the treatment? Are there any cancer risks following the consumption of irradiated food? Does irradiated food become radioactive? Does irradiation adversely affect the bacterial population in the products? Does it work as a method of preservation?

The vitamin losses in irradiated foods vary, on average, from 15% to 70% depending on the food and the nutrient.When irradiated food is stored for eight months it loses twice the amount of nutrients as ordinary food. During irradiation, fatty foods, especially polyunsaturated oils go rancid, while irradiating carbohydrate foods can produce formaldehyde, an undesirable toxin. There is some evidence that *aflatoxin* production, another undesirable, increases up to 50 times in irradiated foods.The only good news for irradiation is that it does destroy undesirable bacteria, it does not make the food radioactive, nor, based on the very limited research so far available, does it appear to affect human health. However, it appears to have little effect in increasing the shelf life of fruit.

Since there have already been incidences of sub-standard contaminated foods being irradiated to destroy pathogenic bacteria, and then being resold it is dubious that the introduction of irradiated food will benefit the consumer even if it does benefit the food producers.

ANTI-BIOTIC RESIDUES

Approximately half the anti-biotics produced are used in low levels in animal feed to increase growth rate and avoid infection. This has resulted in some bacteria becoming resistant to antibiotics. While there is no evidence yet that these bacteria are passed on to meat eaters, there is evidence that people who handle raw meat or feed containing antibiotics have a large number of resistant bacteria in their gut. In view of the possibilities it is wise to eat organic or free-range animal produce.

NATURALLY OCCURRING TOXINS

Although toxic substances are not good for you it is an illusion to think that you can avoid them all. Many natural foods contain small amounts of toxic substances. Some of these protect the plants and as such are natural pesticides.

Some foods contain *amines*, for example *tyramine*, which interfere with neurotransmitters, chemicals involved in nerve transmission. Tyramine rich foods can induce headaches. These foods include various cheeses, yeast extracts, pickled herring, meats, sausages and wine. Chocolate contains a substance called *phenylethylamine* as well as *caffeine* which may raise blood pressure and precipitate a headache.

Caffeine is a particularly powerful amine. It is found in coffee, tea, cola drinks and chocolate. As well as being a highly addictive stimulant it can induce irregular heartbeat, 'restless legs' syndrome, nausea, headaches and even vomiting in sensitive individuals. Recently there has been increasing evidence that the risk of pancreatic and perhaps other forms of cancer is higher in those who consume large amounts of caffeine.

Other potentially carcinogenic compounds include *psoralens* founds in parsley, parsnips and celery, *alkaloids* found in the flowers of some herbs and herb teas, and *isothyocyanate* in mustard seeds and horse radish. However these foods also contain anticarcinogenic factors so it is difficult to say what their net effect is. Most naturally occurring carcinogenic factors exert their effect by generating free radicals. Therefore, if your diet is high in anti-oxidant nutrients, namely vitamins A,C,E, zinc and selenium consumption of these foods is less likely to be of concern.

DO YOU NEED DRUGS?

As a nation , our consumption of medical drugs is the highest in the world. The British pharmaceutical industry turns over a staggering £2,000 million a year. But do we really need to take so many? The answer is no.

For example, the use of aspirin for mild pain and headaches can never do us any long term good. The human body has no need for its active ingredient *salicylic acid*. Continual use of aspirin is known to increase risk of stomach ulceration and kidney disease, as well as blocking vitamin C uptake and lowering folic acid levels. Some people take pain killers containing both aspirin and caffeine, since coffee drinkers deprived of coffee will suffer headaches as a sign of withdrawal. Out of the frying pan into the fire!

Another common and extremely dangerous group of substances are cortisone-based drugs. *Cortisone* is a synthetic form of the hormone produced by the adrenal gland to combat stress. This substance has almost magical qualities in that it can reduce pain, stop inflammatory reactions, prevent transplanted organs being rejected, and is now being used to treat over 100 different ailments including cancer, arthritis, kidney disease, hay fever and allergies. In America 29 million prescriptions for it are written each year! "The sad truth is that cortisone doesn't cure anything: it merely suppresses the symptoms of the disease" says Dr. Zumoff, formerly with the Steroid Research Laboratory at New York's Montefiore Hospital. One of the major problems with this class of drug is that the body stops producing its own cortisone. Wherever possible drugs are best avoided.

 **AVOID**

- Avoid all food additives except those listed on page 33.
- Avoid foods packaged in aluminium.
- Don't cook your food in aluminium or copper pans.
- Buy milk in bottles rather than cartons.
- Buy teas from companies who use dioxin free tea bags.
- Avoid irradiated food.
- Eat only organic or free-range meat, chicken or eggs.
- Limit your intake of coffee, chocolate, and alcohol.
- Fry as little as possible.
- Avoid excessive use of pain killers.
- Avoid antacids containing aluminium.
- Avoid frequent use of anti-biotics.
- Avoid anti-inflammatory drugs.
- Avoid the use of anti-depressants and sleeping pills.
- Investigate natural and barrier methods of birth control rather than taking the pill.

Do not stop or reduce any prescribed medication without first consulting your doctor.

 PROTECT

- Since foods without preservatives are more likely to go off, it's important to buy fresh produce and consume it relatively quickly.
- Keep cold-pressed oils in the fridge.
- If you take pain killers regularly take at least 1,000mg of vitamin C daily.
- If you are on anti-biotics take a high strength B complex during the course, and supplement beneficial bacteria for an additional two weeks to recolonise your gut with good bacteria.
- If you are on the pill take a high strength B complex and extra B6 to provide 100mg a day, plus zinc 15mg.
- If you take sleeping pills or anti-depressants also take a yeast-free high strength B Complex.

CHAPTER TWO

HOW POLLUTED ARE YOU?

Most pollutants damage the body in similar ways. Therefore, it's just as important to keep your overall level of pollution down as it is to avoid a particular pollutant. Long term exposure to low levels of pollution can be just as bad as a short term high level exposure. It is for these reasons that it has taken us so long to realise that escalating world pollution does pose a serious health hazard.

Pollution is also far more dangerous when your defences are down, at times of stress or rapid growth. So these tips are especially important for children and teenagers, pregnant or breastfeeding women and people with immune or respiratory diseases.

The questionnaire overleaf allows you to identify sources of potential pollution which helps you to pinpoint what you need to do to decrease your overall risk and protect yourself against the many unavoidable hazards of modern living.

Tick each box if you answer YES.

HOW TO PROTECT YOURSELF FROM POLLUTION

LEAD, ALUMINIUM AND OTHER TOXIC METALS

☐ Do you spend more than two hours a week in heavy traffic?
☐ Do you buy unwrapped fruit or vegetables from stalls ex
  posed to street traffic?
☐ Do you eat coastal fish or shellfish?
☐ Do you drink more than one measure of alcohol a day?
☐ Do you live in a soft water area?
☐ Do you cook with aluminium or copper saucepans?
☐ Do you grill foods on or wrap foods in aluminium foil?
☐ Do you have five or more amalgam fillings?
☐ Do you have lead water pipes? (You probably do if your house
  is pre 1940.)
☐ Do you live in a fluoridated area?

NITRATES AND PESTICIDES

☐ Do you eat cured meat e.g. ham, sausages, bacon, pies?
☐ Do you drink tap water rather than bottled or filtered water?
☐ Do you generally eat non-organic produce?
☐ Do you wash or peel non-organic fruit and vegetables?
☐ Do you eat fruit and vegetables that are not in season?
☐ Do you regularly eat a lot of protein (meat, fish, cheese, eggs)?
☐ Do you eat (non-organic) meat most days?
☐ Do you have difficulty digesting protein-rich foods?
☐ Do you only brush or floss your teeth occasionally?
☐ Do you live in an arable area?

FREE RADICALS, RADIATION AND AIR POLLUTION

☐ Do you smoke or work in a smoky atmosphere?
☐ Do you have excessive exposure to strong sunlight,or
  sunbeds?
☐ Do you often eat fried foods?
☐ Do you eat smoked or barbecued foods?

HOW POLLUTED ARE YOU ?

☐ Do you exercise near busy roads?
☐ Do you use aerosol sprays?
☐ Do you live in an inner city?
☐ If so, do you have neither an air filtration system or ioniser at work or at home?
☐ Do you spend a considerable amount of time in front of a VDU screen?
☐ Do you suspect you live in a high radon area or are exposed to high levels of electromagnetic or other radiation e.g. by an electricity pylon, radio, radar or nuclear power station?

FOOD ADDITIVES AND DRUGS

☐ Do you occasionally take antacids?
☐ Are you on the contraceptive pill or hormone replacement therapy?
☐ Do you regularly take painkillers e.g. aspirin?
☐ Do you use anti-biotics twice or more a year?
☐ Do you take anti-inflammatory drugs?
☐ Do you take anti-depressants or sleeping pills?
☐ Do you often eat prepackaged foods?
☐ Do you often eat fast foods?
☐ Do you eat foods that contain colourings or additives?
☐ Do you drink milk out of cartons rather than bottles?

SCORING YOUR POLLUTION RISK

Each section of this questionnaire relates to a potential source of pollution. Score one point for each Yes answer, the maximum score for each section being 10. The higher the score in one area, the more important this is for you. Any score above five means that this area warrants your serious attention. Add up your score for each section to arrive at your grand total. Then shade in each square of the 'toxic barrel' for each YES answer to see whether you fall in the high, medium or low pollution group.

If you scored.....

30-40 Your risk of pollution is too high. Read each section carefully and see what you can do to change your diet and lifestyle. Supplement your diet with anti-oxidant nutrients, plus extra vitamin C, calcium and zinc. These will help you to detoxify.

20-30 Your pollution risk is quite high. Its time to take action and revamp your diet and lifestyle. Pay particular attention to those sections you scored four or more on. Write down the things you could do to lower your score and stick to them strictly for one month.

10-20 You're better than average, but there's still room for improvement. Write down the things you could do to lower your score and stick to them for a month. If you scored four or more on any section this warrants your attention.

0-10 Well done. You're already pollution conscious and are unlikely to be affected by pollution. Now its time to fine tune by looking at those questions you answered YES to - and can change. If you scored four or more on any one section this warrants your attention.

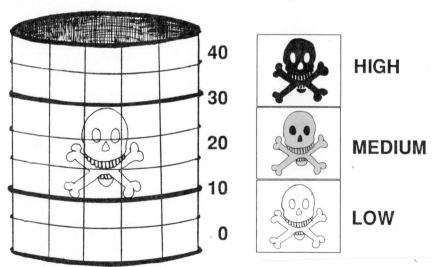

40

30

20

10

0

HIGH

MEDIUM

LOW

CHAPTER THREE

THE POLLUTION PROTECTION PLAN

As well as avoiding obvious sources of pollution in your environment you can eat for protection. The first step is to choose foods free of pollutants. That means becoming aware of how foods are packed, cooked, prepared, processed and grown. If you don't want foods packaged in aluminium don't buy them. If you want organic produce at reasonable prices tell your green grocer. Food producers will soon get the message. Compulsory labelling for food additives was introduced in 1986. Already there are so many foods available without additives.

The second step is to choose foods that contain positive factors that actually protect you against pollution. The aim of a detoxifying diet is to provide: foods rich in nutrients that disarm pollutants and strengthen your body's defence mechanisms; foods rich in nutrients that strengthen the surfaces through which pollutants enter, namely your digestive tract, lungs and skin; foods that help you to detoxify; foods that cleanse your digestive tract and help you to digest and absorb food properly.

THE DETOXIFYING DIET

Here are ten simple tips to follow every day to ensure that you are well protected from pollution.

1. Eat at least three pieces of washed fresh fruit (preferably organic). Bananas and apples are the best.
2. Have something green or something raw with every meal.
3. Have a carrot, rich in vitamin A, every day.
4. Have a tablespoonful of sesame seeds, sunflower seeds or a dozen almonds. These foods are rich in calcium, magnesium,zinc and vitamin E.
5. Eat plenty of onions, garlic or the occasional egg since these foods are high in the sulphur containing amino acids which help you detoxify.
6. Eat foods as unadulterated as possible i.e. wholegrains, fresh fruit, unprocessed foods.
7. Limit your intake of alcohol, coffee, tea, sugar and refined foods.
8. Limit your intake of non-organic or non-free range meat to a maximum of three times a week.
9. Drink a pint of bottled or filtered water or diluted fruit juices.
10.Take an anti-oxidant supplement.

SUPPLEMENTS TO STRENGTHEN YOUR DEFENCES

Most pollutants are anti-nutrients, that is they interfere with, prevent the absorption of, or promote the excretion of vital nutrients. As well as avoiding sources of pollution you can protect yourself by ensuring an ideal intake of key vitamins, minerals and amino acids. Not only does this prevent any potential losses due to pollutants, but it also keeps your detoxifying mechanisms at full strength. Some people like to think that you can maintain maximum protection against pollution simply by eating a good diet. This simply isn't true. You need extra protection against the onslaught of heavy metals, nitrates, pesticides, free radicals, radiation and all the other unavoidable sources of pollution.

POLLUTION PROTECTION PLAN

Nutrients help protect against pollution in three ways:

1. Antagonists of undesirable elements - calcium, magnesium, zinc and selenium are antagonistic to lead, cadmium, aluminium, mercury, fluorine, chlorine and arsenic. They may prevent their uptake or stop them doing damage. By making sure that your enzyme systems have more than enough of the right mineral co-factors these rogues have a hard time taking their place.

2. Anti-oxidants - vitamin A, (both in the form of retinol, the animal source, and beta-carotene, the vegetable source) vitamin C and vitamin E are all anti-oxidants. They help to disarm free radicals, which means protection against air pollution, smoke, radiation and rancid foods. The amino acids l-cysteine, l-glutathione, and the minerals zinc, selenium, and to a lesser extent, iron and manganese, are components of anti-oxidant enzymes which also help to disarm free radicals. A special detoxifying system in the liver also requires these nutrients to break down persistent chemicals such as pesticides into substances that can be eliminated. Vitamin C also prevents the formation of nitrosamines from nitrites.

3. Natural chelators - pectin, found mainly in fruits including bananas, apples, but also carrots, and vitamin C are called *chelators* (from the Greek word 'chela' meaning a claw) because they can latch onto heavy metals and remove them.

The ideal intake of these nutrients to take for 'pollution protection' is shown in the table below. If you have reason to believe you are, or have been exposed to a considerable amount of pollution you may wish to take the levels indicated for 'detoxification'. If you take these levels daily for three to six months, as well as following the Detoxifying Diet and avoiding possible sources of pollution, you should significantly lower body levels of pollutants and strengthen your defences. It usually takes at least three months to significantly lower body levels of heavy metals.

IDEAL INTAKE OF NUTRIENTS FOR POLLUTION PROTECTION AND DETOXIFICATION

Nutrient	For Protection	For Detoxification
Vitamin A*	7,500iu	15,000iu
Vitamin C	1,000mg	3,000mg
Vitamin E**	100iu	300iu
Calcium	300mg	600mg
Magnesium	150mg	300mg
Iron	6mg	12mg
Zinc	10mg	20mg
Manganese	2.5mg	5mg
Selenium	25mcg	75mcg
L-cysteine	100mg	200mg
L-glutathione	25mg	75mg
Pectin	100mg	300mg

*Vitamin A is best supplied in the form of beta-carotene.
*Vitamin E is best supplied as d-alpha tocopherol.

Many vitamin companies produce supplements that contain a combination or all of these key nutrients. It is therefore best to shop around and find a formula that meets your needs. In addition to supplementing these nutrients it is wise to take a general high strength multivitamin and mineral supplement. These may also contain significant amounts of the key nutrients discussed here.

In the chart opposite we have compared the four best supplement formulas currently available, from a selection of twelve 'anti-oxidant' products.

POLLUTION PROTECTION PLAN

Supplement	Vit A	Vit C	Vit E	Zinc	Sele-nium	Calc-ium	Magne sium	L-Cyst-eine	L-Gluta-thione
For Protection	7,500iu	1,000mg	100iu	10mg	25mcg	300mg	150mg	100mg	25mg
For Detoxification	15,000iu	3,000mg	300iu	20mg	75mcg	600mg	300mg	200mg	50mg
Health+Plus Detox Formula II	7,500	1,000	100	10	25	300	150	100	25
Solgar Anti-Oxidant Factor	11,000	600	250	22	75	-	-	100	25
Nature's Best Beta-Max Carotene Complex	8,325	1,000	100	-	25	-	-	-	-
Wassen Selenium ACE	1,500	90	45	-	100	-	-	-	-

Solgar's Anti Oxidant Factor and *Wassen's* Selenium ACE are available widely in health food shops. *Health+Plus'* Detox Formula II is available by mail order from Health+Plus Ltd 118 Station Road, Chinnor OX9 4EZ (Tel:0844 52098). *Nature's Best's* Beta-Max Carotene Complex is available by mail order from Nature's Best Ltd, Freepost PO Box 1, Tunbridge Wells TN2 3EQ (Tel:0892 34143).

POLLUTION PROTECTION FOR THE PLANET

The book will help you clean up your inner environment. But what about the environment we all share? Imagine how our world has changed in the last fifty years. The air is becoming unbreathable, the water undrinkable, the soil is seriously poisoned. Trees are being destroyed. Wildlife is becoming extinct. Do you think we have the reserves to survive yet more man-made and natural disasters? Do you really think our future is guaranteed fifty years from now?

If we all act now, even in small ways, we will solve the problems of pollution. If we wait, it will soon be too late.

RECOMMENDED READING

Patrick Holford - *The Family Nutrition Workbook* (Thorsons) 1988.
Karen Christensen - *Home Ecology* (Arlington Books) 1989.
John Button - *How To Be Green* (Century) 1989.
Robert Buist - *Food Chemical Sensitivity* (Prism Press) 1986.
Nigel Dudley - *Nitrates - The Threat to Food and Water* (Green Print) 1990.
Drs Passwater and Cranton - *Trace Elements, Hair Analysis and Nutrition* (Keats) 1983.

USEFUL ADDRESSES

THE INSTITUTE FOR OPTIMUM NUTRITION (ION) offers courses and personal consultations with trained nutritionists. To receive ION's information pack please ring or write to ION, 5 Jerdan Place, London SW6 1BE. Tel. 071-385-7984

FRIENDS OF THE EARTH is Britain's best-known environmental action group. You can join Friends of the Earth by contacting them at 26-28 Underwood St., London N1 7JQ. Tel 071-490-1555.

MORE HEALTHY READING FROM ION PRESS

HOW TO BOOST YOUR IMMUNE SYSTEM
by Jennifer Meek
This book explains how the immune system works and what you can do to make yours healthy. ISBN 1-870976-00-2 (£2.50)

THE ENERGY EQUATION
by Patrick Holford
Energy, or rather a lack of it, is the most common sign of ill-health. But what is energy, how do you make it and get more of it. ISBN 1 87097-601-0 (£1.99)

SUPERNUTRITION FOR A HEALTHY HEART
by Patrick Holford
This book explains what causes heart disease and how to prevent it with optimum nutrition. ISBN 1-870976-02-9 (£1.99)

HOW TO IMPROVE YOUR DIGESTION AND ABSORPTION
by Christopher Scarfe
You are not what you eat. You are what you can digest and absorb. This book explains how to get the most out of your food. ISBN 1-870976-03-7 (£1.99)

These books can be ordered from any bookshop, or direct from
ION Press, 5 Jerdan Place, London SW6 1BE (Tel:071 385 7984)